Detroit's Sports Broadcasters

On the Air

DETROIT'S SPORTS BROADCASTERS
On the Air

George B. Eichorn
with an introduction
by Ernie Harwell

ISBN 0-7385-3166-0

Published by Arcadia Publishing,
an imprint of Tempus Publishing, Inc.
Charleston SC, Chicago, Portsmouth NH,
San Francisco

Printed in Great Britain.

Library of Congress Catalog Card Number: 2003108761

For all general information contact Arcadia Publishing at:
Telephone 843-853-2070
Fax 843-853-0044
E-Mail sales@arcadiapublishing.com
For customer service and orders:
Toll-Free 1-888-313-2665

Visit us on the internet at http://www.arcadiapublishing.com

For Sue, Dianna, and Erica—

Your sacrifices are immeasurable yet so greatly appreciated.

CONTENTS

ACKNOWLEDGMENTS

Heart-felt gratitude to the following individuals who contributed time, photographs and encouragement to this book: Isabella Agby, Bob Benko, George Blaha, Tim Bryant, Bridget Burns, Paul Carey, Bill Eisner, Ben Eriksson, Dave Frisco, John Fountain, Dan Graschuck of El Central, Jennifer Hammond, Ernie Harwell, Margaret Hehr, Mark A. Hicks, Jim Hindy, Bill Janitz, Bob Kiess, Ray Lane, Lindy Lindell, Laura Loviska, Budd Lynch, Horst Mann of The Detroit Monitor, Scott Morganroth of The Detroit Monitor, Mark Pattison, Paul Pentecost, Fred Pletsch, David Poremba of the Burton Historical Collection (Detroit Public Library), Dave Raglin, Jack Riggs, Tom Ryan, Ellen Sawyer, Tom Ufer, Jeff Weiss, Robert L. Wimmer, and all others whose photos are credited herein.

INTRODUCTION

by Ernie Harwell

The perfect selection to write this inside look at the Detroit sports media is George Eichorn. And nobody could have done it better.

George has been a part of the local sports scene since the 1970s. A longtime observer, he himself also has participated as both a writer and radio performer. In addition, he has seen the other side of the sports picture as a public relations expert for various organizations.

In his capacity as executive administrator of the Detroit Sports Broadcasters Association, Eichorn is in close touch with the doings of these reporters. He knows them all, and has known them since most of them made their debuts. Eichorn wasn't around when Ty Tyson pioneered radio play-by-play of football and baseball. Granted, George never knew Ty, but he knew about him. As for the others who came along after Tyson, George had a personal relationship with many of them.

Eichorn narrates a great parade of talented performers in *Detroit Sports Broadcasters: On the Air*. After Tyson, along came such stars as Harry Heilmann, Harry Wisner, Van Patrick, David Diles, Budd Lynch, Bob Ufer, Bill Fleming, Bob Reynolds, Paul Carey, George Kell, Ray Lane, Al Ackerman, Frank Beckmann, Don Shane, George Blaha, and Dan Dickerson.

In this well-written book, we get inside glimpses of these stars—some of them nationally known. Also, the photos are outstanding and put our on-the-air heroes on display—many times in informal settings.

Detroit has always been a great sports town and a top market for sports radio and television. Thank's to Eichorn's insightfulness, we have a close-up look at the broadcasters who made sports in Detroit come alive.

Relax, read, and enjoy George Eichorn's insight into a fascinating world.

ONE
Baseball

Edwin "Ty" Tyson is the president in perpetuity of the Detroit Sports Broadcasters Association and a sports broadcaster of legendary proportions. He broadcast Detroit Tigers games—with great articulation and a great wit—from 1927 to 1942, as shown here at Briggs Stadium in downtown Detroit. Tyson also did University of Michigan football games on radio and holds the distinction of being behind the microphone for the first live play-by-play coverage ever for a sports event when he called the 1924 Michigan/Wisconsin game at Ferry Field in Ann Arbor. Ty was inducted in 1996 into the Michigan Sports Hall of Fame. (From the Michigan Sports Hall of Fame.)

Harry Heilmann played for the Detroit Tigers from 1914 to 1929 and was named to the All-time Tiger Team for the first 50 years of their participation in the American League. This outfielder par excellence was a logical choice by Tigers management to become the team's broadcaster in 1934, on WXYZ-AM. He concluded his Tigers broadcast career in 1950. Heilmann was inducted into the Baseball Hall of Fame in 1952 and Michigan Sports Hall of Fame in 1956. "Listen to the voice of baseball" and "Bugaboo! Another fly is dead!" were Heilmann trademarks. (From the Burton Historical Collection.)

Paul Williams broadcast Tiger baseball from 1947 to 1952 on radio and television, working with legends Ty Tyson and Harry Heilmann. Williams was also a sportscaster for several Detroit radio stations including WWJ-AM and WWJ-TV. He was Detroit Sports Broadcasters Association president from 1950–51. (From the Burton Historical Collection.)

William Ernest Harwell has been the voice of summer almost every year since 1960, calling Detroit Tigers games on radio or television. Ernie once had his tennis shoe autographed by Yankees legend Babe Ruth at an exhibition game at Ponce de Leon Park in Georgia. He never saved the shoe, but Ruth got a kick out of Ernie's determination to secure The Babe's autograph despite not carrying a paper to sign it on! This undated photograph shows Ernie around the time of his calling play-by-play for the New York Giants and Brooklyn Dodgers in the 1940s and 1950s. (From the Burton Historical Collection.)

George Kell (left) and Ernie Harwell formed the Detroit Tigers broadcast team from 1960-63 as shown here. Kell was an all-star third basemen with the Tigers before moving into the broadcast booth. Harwell joined the Tigers after calling games for the Baltimore Orioles. Both men are members of the Baseball Hall of Fame and Michigan Sports Hall of Fame. (From the Burton Historical Collection.)

The "Year of the Tiger" was 1968. Ernie Harwell (right) is flanked by broadcaster Ray Lane (center, facing camera) in the old broadcast booth at Tiger Stadium, complete with chicken wire to prevent foul balls from propelling into the pair. Harwell and Lane worked together from 1967 to 1972. (Collection of Ray Lane.)

"Move Over Babe (Here Comes Henry)" was a 45 record written by Ernie Harwell (left) and recorded by Detroit Tigers pitcher Bill Slayback, capturing the magical journey of Atlanta Braves slugger Henry Aaron (right) in overtaking all-time major league home run king Babe Ruth. Ernie presented Aaron with a copy of the single in 1973. (From the Burton Historical Collection, and from Karen Records.)

Paul Carey (right) is shown interviewing baseball great Ted Williams before a spring exhibition game in the mid-1970s in Florida. Carey was the appropriate successor to Ray Lane on Tigers radio broadcasts in 1974, with some 20 years of broadcast experience at WJR-AM. Paul would spend 19 seasons in the booth with Ernie. (Collection of Paul Carey.)

Whether calling a game or visiting the fans in the crowded corridors of Tiger Stadium, Harwell (left) and Carey were immensely popular. Their friendship continues today, long after Carey retired from active broadcasting (From Glenn L. Hibbert.)

Joe Gentile (left) flanks Ernie Harwell and Paul Carey in the mid-1980s. Gentile was president of the Detroit Sports Broadcasters Association from 1962 to 1964. His popular "Happy Joe's Early Morning Frolic" on CKLW-AM was co-hosted by Ralph Binge. The two dominated Detroit's morning-drive time with antics and anecdotes. Gentile went on to become public address announcer at Tiger Stadium for many years. (Collection of Paul Carey.)

Sparky Anderson (left) converses with Tigers television "voice" George Kell at one of owner John Fetzer's January press parties at the Detroit Athletic Club. Kell formed a bond with the popular Tigers manager and the viewing public. Kell sayings included: "He hit it a mile!," "We have a confab out on the mound," and "High skies here at Tiger Stadium." (From Charles Jackson.)

Ernie Harwell (left) joins Tigers manager Sparky Anderson at the debut of the "IBM Sports Gallery" in the 1980s at the Detroit Historical Museum. Harwell's career as a Tigers broadcaster spanned 18 managers (including interim ones). Like Harwell, Anderson is a member of the Baseball Hall of Fame. Anderson has broadcast experience with CBS Radio and the Anaheim Angels television network.

One of the most controversial moves in the history of Detroit sports broadcasting was the decision by Tigers management and WJR-AM to replace Ernie Harwell in the radio booth following the 1991 season, with Rick Rizzs (left) and Bob Rathbun (right). The pair never caught on with the public or the media and were united with Harwell for one season—1993—as shown here, after Mike Ilitch purchased the Tigers from Tom Monaghan (who allowed the Harwell dismissal). Harwell spent 1994-98 as a Tigers telecaster but returned to the radio booth in 1999. (From Michigan Radio Guide.)

The final season of baseball played at "The Corner" of Michigan and Trumbull—Tiger Stadium—brought this reunion of baseball broadcasters. From left to right are Ray Lane, Ernie Harwell, Paul Carey, George Kell, Al Kaline, Frank Beckmann, and Larry Osterman. They were

honored in special pre-game ceremonies prior to the ballpark's closure. Beckmann is still broadcasting Tigers games on WKBD-TV, at new Comerica Park. (From Bill Eisner.)

Detroit Tigers and Fox Sports Net broadcaster Josh Lewin (left) is honored as the first recipient of the "Ty Tyson Excellence in Sports Broadcasting" Award from the Detroit Sports Broadcasters Association, in June of 2001. Presenting the plaque is D.S.B.A. President Ken Kal at the Southfield Westin Hotel. Lewin worked for FSN in Detroit from 1998-2001 before moving south to call the Texas Rangers games. Lewin also calls Saturday Game of the Week baseball contests for Fox Sports/national. (From Bob Benko.)

Shown here and the next page receiving the 2002 Detroit Sports Broadcasters Association Ty Tyson Award from D.S.B.A. officers are, from left to right, George Eichorn, Jim Stark, and Bill Harrington. Harwell called his final season of Tigers baseball in 2002, on flagship station, WXYT-AM. The Georgia native called Bobby Thompson's "Shot heard 'round the world" home run in 1951 during the first national coast-to-coast telecast of a major sports event. (From Ben Eriksson & Laura J. Loviska.)

To George
Ernie Harwell
DETROIT SPORTS BROADCASTERS
Ty Tyson
AWARD
"Excellence In Sports Broadcasting"
2001
Ernie Harwell

Ernie Harwell's career is captured in the 2002 Detroit Tigers official scorecard, which sold for $1.00 at Comerica Park that season. Ernie was showered with gifts and tributes from the Tigers and at every major league stadium he called a final game in. (From the Detroit Tigers.)

Mario Impemba (left) succeeded Josh Lewin as Fox Sports Net's Tigers play-by-play announcer in 2002. He is shown here with actor/director Jeff Daniels (center) and FSN analyst Kirk Gibson (right). Impemba is a Detroit native who joined FSN after seven successful seasons as the radio and fill-in television play-by-play man for the Anaheim Angels. The Tucson Toros, Quad City Angels, and Peoria Chiefs also dot Impemba's resume. (From Fox Sports Net.)

Dan Dickerson (left) and Jim Price form the Tigers' radio broadcast team in 2003. Dickerson made his Tigers broadcasting debut in 2000 and gained valuable experience working alongside Ernie Harwell. He's worked for WJR-AM, WWJ-AM, and now WXYT-AM as co-host of the *Tigertown* call-in show. Price was a member of the 1968 World Champion Tigers and has been part of Tigers broadcasts and telecasts dating back to 1993. (From Ben Eriksson.)

Frank Beckmann (left) teams with former Detroit Tigers pitcher Jack Morris (right) on 40 Detroit Tigers telecasts on UPN-50 (WKBD-TV). Beckmann's career spans 32 years on WJR-AM and several on UPN-50. He began as a new reporter for WJR in 1972 and won an award for his coverage of the Jimmy Hoffa disappearance. His love of sports got the best of him and he started the *Sportswrap* call-in show. Beckmann became radio "voice" of the Tigers in 1995 until '98. The three-time Michigan Sportscaster of the Year is with his third full-time partner on Tigers telecasts (Al Kaline and Lance Parrish were the others). Morris pitched for the Tigers from 1977-90, throwing a no-hitter during the 1984 World Championship season. (From Ben Eriksson.)

Mario Impemba (left) is joined in 2003 by former Detroit Tigers outfielder Rod Allen on Fox Sports Net. Allen spent five seasons as television and radio analyst on Arizona Diamondbacks games before coming to Detroit. In addition, Allen has worked Fox Saturday Baseball regional telecasts and the National League Division Series. Rod played on the Tigers' 1984 World Championship team but was drafted by the Chicago White Sox. His son, Rod Allen Jr., is a top baseball prospect from Arizona State University.(From Fox Sports Net.)

Al Kaline has been dubbed "Mr. Tiger"—and why not? He was the longtime color analyst on Detroit Tigers telecasts, working with George Kell. Kaline has 50 seasons with the Tigers as a player, broadcaster, spring training coach, and now as assistant to the president. He may not be the most polished broadcaster but Kaline never shyed away from being insightful. (From Ben Eriksson.)

Detroit native Tom Paciorek broke into the majors leagues in 1970 with the Los Angeles Dodgers and went on to play 18 seasons, averaging .282. He concluded his career in 1987 with the Texas Rangers. Paciorek also played for the Atlanta Braves, Seattle Mariners, Chicago White Sox and the New York Mets. He moved into the broadcast booth with the White Sox where he teamed for many seasons with Ken Harrelson. Paciorek also has filled in on Fox Sports Net for Detroit Tigers telecasts and now does Atlanta Braves games for Fox Sports Net South. Tom was inducted into the National Polish-American Sports Hall of Fame in 1992. Tom's brothers, John and Jim, also played in the major leagues.

Two
Football

Van Patrick (right) is synonymous with football broadcasting, as he once was with Tigers baseball. Dubbed "The 'Ol Announcer" by his many fans and media colleagues, Van was the radio "voice" of the Detroit Lions, Tigers, and Notre Dame football. His Tigers' stint began in 1951 and lasted 10 seasons, before Ernie Harwell and George Kell became a team. Patrick was in the Lions' radio booth from 1950 until his untimely death in 1974. The Texas native was colorful in every way with phrases like, "He's at the 30! He's at the 20! He's at the 10! He's at the 5...he's in for a touchdown!" Patrick is shown here with Paul Pentecost, his longtime spotter, in a visit to Hazel Park Raceway in 1965. (Collection of Paul Pentecost.)

In this 1966 photo taken at the Los Angeles Coliseum, Detroit Lions radio broadcasters Ray Lane (left) and Bob Reynolds (center) flank spotter/statistician Walter Dell. Reynolds teamed with Van Patrick most games in this era. Check out the old-style attached microphone around Reynolds' neck and the blazers they are wearing! (Collection of Ray Lane.)

Van Patrick anchored sports at WJBK-TV in Detroit and at Mutual Broadcasting System (voice of Notre Dame football) and was part of the Class of 1991 for the Michigan Sports Hall of Fame. He's shown here with Notre Dame Coach Ara Parseghian at an unidentified location and date. (Collection of Paul Pentecost.)

George Allen (left), seen in this 1981 photo with legendary announcer Lindsey Nelson (right) was born and raised in Michigan. The former head coach of the Los Angeles Rams and Washington Redskins, Allen was a successful pro football analyst for CBS in the 1980s. He was inducted into the Pro Football Hall of Fame in 2002 as his son, George, accepted the award on behalf of his late father. Allen loved coaching and broadcasting football. (From CBS Sports.)

Rick Forzano coached the Detroit Lions from 1974 to 1976, going 15-18, before being replaced by Tommy Hudspeth four games into the '76 season. Rick went on to become an ABC-TV college football analyst, doing regional games. He still makes his home in the Detroit area and is an occasional dinner speaker. (From Detroit Lions.)

In this 1983 CBS Sports group photo, former Detroit Lions great Wayne Walker is among a team of NFL broadcasters. He's at the far left in the middle row. Walker, a member of the Michigan Sports Hall of Fame, played in Detroit from 1958 to 1972, and immediately started a broadcasting career at WJBK-TV, as part of its "Five Star Sports" lineup that included Lions broadcasters Van Patrick and Ray Lane. Walker later became an accomplished TV sports anchor in San Francisco before retiring in 2000. (From CBS Sports.)

Dick Enberg has his roots in Michigan. A native of Armada, he was a standout aspiring young sports and news broadcaster at Central Michigan University (CMU) in Mount Pleasant, and was elected student body president in 1957. Enberg rose to prominence in 1965 when he earned the main broadcast position with baseball's California Angels. Enberg later called UCLA basketball and Los Angeles Rams football before being tapped by NBC as its lead announcer for baseball, football, and the Olympic Games. A member of the CMU Athletic Hall of Fame and Michigan Sports Hall of Fame, Enberg now works for CBS Sports.

Bob Ufer was the voice of "MEECHIGAN" football for five decades in a career that spanned 363 consecutive games from 1945 to 1981. He also was a world-class athlete in his own right, playing football his freshman year in the same backfield as Heisman Trophy winner Tom Harmon (also a Michigan broadcaster). Ufer made his mark in track where he held every Michigan record from the 100 to 800 yard dash. His memory lives on through scholarships given annually in his name to student athletes going to the U of M. Football Coach Lloyd Carr says, "The Michigan football tradition is all about this kind of enthusiasm and energy. Bob's broadcasts will never be duplicated!" Bob was elected to the Michigan Sports Hall of Fame in 1992. (From Tom Ufer.)

Two sportscasters with Michigan ties were part of CBS Sports' NCAA Football broadcasting class of 1983. Standing at far left is Charlie Neal, and Steve Grote stands fifth from left (with mustache). Neal also was a Detroit sports anchor/reporter on WJBK-TV and WJR-AM during the 1980s. Grote was a University of Michigan basketball star who made the transition into broadcasting after his playing career. (From CBS Sports.)

Former all-pro Ron Kramer (right) worked along with TBS sports announcer Pete Van Wieren on TBS' package of syndicated Big Ten Conference college football games in the 1980s. Kramer made his mark as an All-American with the University of Michigan Wolverines (1954-56) before signing a professional contract with the Green Bay Packers (1957). He starred with the Pack for eight NFL seasons before concluding his pro career with the Lions (1965-67). Kramer was elected to the Michigan Sports Hall of Fame in 1971. (From TBS Sports.)

Dennis Franklin guided the University of Michigan to a 30-2-1 record as quarterback, yet never played in a post-season game with the Wolverines. He played briefly in the NFL and CFL, and later became an expert analyst on broadcasts of NCAA Football on WWJ-AM, ABC-TV, and CBS-TV. In 1982, Franklin joined CBS Sports' stable of gridiron analysts that included Pat Haden, Brian Dowling and Steve Davis. (From CBS Sports.)

Michigan Panthers games in 1984 were carried by WXYZ Radio and 14 stations across Michigan. Veteran, award-winning sportscaster Bob Sherman (right) was the voice of the Panthers of the United States Football League (USFL). His color analyst was Dan Follis (left), a longtime Michigan State supporter. Sherman's credits included play-by-play experience with Michigan State University, Ohio State University, and the Indianapolis 500. (From Michigan Panthers.)

Jim Brandstatter (left) played three varsity years for Glenn "Bo" Schembechler (right) at the University of Michigan and later got to interview the coach while working for WJR-AM. During Jim's playing career and later, "Brandy" would forge a lasting friendship with Schembechler. In 1980, Brandstatter started co-hosting the "Michigan Replay" coach's show with Bo, and today the two have a seasonal "Brandy And Bo" show on WJR. (From WJR-AM.)

In this 1987 photo, Dan Dierdorf (left) is partnered with Frank Gifford and Al Michaels for ABC-TV's *Monday Night Football*. Dierdorf was an All-American offensive tackle while playing at the University of Michigan for Bo Schembechler, and later became an accomplished football analyst. He has moved on to CBS Sports doing NFL games. Dierdorf was inducted into the College Football Hall of Fame in 2000 and the Michigan Sports Hall of Fame in 2001. (From ABC Sports.)

Michigan Replay with University of Michigan Head Coach Lloyd Carr (left) and co-host Jim Brandstatter is seen statewide in Michigan. The two, together with the production staff, work late Saturdays (following home and away U-M games) in taping the show for its Sunday morning airing. Then Brandstatter must get to his Detroit Lions radio network assignment Sunday morning! (From Michigan Replay.)

Chris Spielman is fast making a name for himself behind the microphone as he did on the football field. The former Detroit Lions linebacker (wearing No. 54) was known for his tenacity and enthusiasm as a player. Now, he's bringing that to the broadcast booth as an ESPN college football analyst. (From Detroit Lions.)

One of Detroit's most recognizable voices is broadcaster Mark Champion (left), voice of the Detroit Lions on WXYT-AM. He also does radio play-by-play for the Detroit Pistons on WDFN-AM. Champion is a Muncie, Indiana native and is famous for his voice-over work with Disney World at the conclusion of every Super Bowl. You hear Champion ask the game's most valuable player, "Where are you going next?" Champion's football calls in Detroit have now been heard for 14 Lions seasons where he works side-by-side with Jim Brandstatter (right). (From David A. Frechette.)

Matt Millen made a career shift from sports broadcasting to NFL team management on Jan. 9, 2001. He left Fox Sports, where he was one of the top pro football analysts, to become president and CEO of the Detroit Lions. His 12-year playing career with the Raiders, Forty-Niners and Redskins was highlighted by three Super Bowl championships, one with each franchise. Millen knows football and now hopes to make the Lions a title contender. (From Dan Graschuck.)

Steve Courtney (left) is sideline reporter on University of Michigan football broadcasts, host of the *WJR Tailgate Show*, and host of *Sportswrap* nightly on the station. Pictured here with U of M play-by-play man Frank Beckmann and color commentator Jim Brandstatter, Courtney also does sports reports on the *Mitch Albom Show* on WJR. Witty and insightful, Courtney brings a refreshing approach to his sports assignments and shows. (From WJR Michigan Radioguide.)

Charlie Sanders was known for his acrobatic and clutch pass receptions as a tight end for the Detroit Lions. He made seven Pro Bowl appearances. After his NFL career ended, Sanders moved into the broadcaster booth as a color analyst for Lions games on WJR-AM, working with Frank Beckmann. Sanders most recently has been a coach and scout for the team he has always loved. (From Detroit Lions.)

Matt Shepard (left) and Rob Rubick host the *Detroit Lions Weekly* program on Fox Sports Net. Formerly a sports anchor at WWJ-AM, Shepard does updates for WDFN-AM, in addition to play-by-play calls of CCHA hockey, Detroit Shock basketball, and Michigan high school football and basketball championships on FSN. Rubick also contributes football commentary to WDFN-AM when not working Lions games for FSN. (From Fox Sports Net.)

Former Detroit Lions players often gain broadcast experience. Greg Landry was the Lions starting quarterback in the 1970s and is the last Detroit signal-caller to earn a Pro Bowl appearance, following the '71 season. Landry worked as a coach for several teams—including the Lions—before becoming a talk show regular on WJR-AM during football Sundays.

Ron Rice played safety for Detroit and, upon his retirement, has gained broadcast experience on WXYT-AM, WWJ-AM and WKBD-TV (Channel 50). His work on Lions pre-game and post-game shows has been insightful for listeners. (From Detroit Lions.)

When he's not busy broadcasting Lions, Fury and Wolverines football, or hosting the Wolverines basketball and football coaches show, Jim Brandstatter (right) fulfills his duties as president of the Detroit Sports Broadcasters Association. In this 1998 ceremony at Tiger Stadium, Tigers shortstop Deivi Cruz accepts his DSBA/Tigers Rookie of the Year Award from Brandstatter.

George Blaha (left) has been the radio voice of Michigan State Spartans football for 25 seasons. Shown here on the links with Earvin "Magic" Johnson, Blaha teamed in 2002 with analyst Larry Bielat and sideline reporter Bill Burke. Blaha and his father attended MSU's first-ever Big Ten game, played at Iowa in 1953. George lists All-American halfback Sherman Lewis and quarterback Charlie Baggett as his all-time favorite Spartan players.

THREE
Hockey

Al Nagler was one of Detroit's most recognizable sports announcers in the 1940s and 1950s. He called play-by-play for a number of seasons for the Detroit Red Wings, preceding Budd Lynch. Nagler also worked for WJBK-TV and radio during and after his hockey broadcasting career. Nagler was president of the Detroit Sports Broadcasters Association from 1951-52. Ironically, Lynch succeeded Nagler in that position too! (From the Burton Historical Collection.)

Budd Lynch is a Detroit and Michigan sportscasting icon. Still going strong as the public address announcer at Joe Louis Arena in Detroit, Lynch has been part of the Red Wings hockey scene for 54 NHL seasons. He started as their radio announcer in 1949 and soon became one of the sport's most colorful announcers, earning a spot in the Hockey Hall of Fame in 1985 as the Foster Hewitt Memorial Award winner. In this 1954 photo, Lynch (left) is shown with his customary cigar alongside Red Wings player Sid Abel, following the team's Stanley Cup Championship victory, at a party in Trenton. (From Leo Cummings.)

Shown here, from left to right, Gordie Howe, Budd Lynch and Maurice "Rocket" Richard participate in a 1961 pre-game ceremony at the historic Montreal Forum. It was the Red Wings Old-Timers versus the Canadiens Old-Timers to benefit charities in the Montreal area and Budd was asked to emcee part of the program. (Collection of Budd Lynch.)

Never shy around a microphone, Budd Lynch (right) stands next to Red Wings Hall of Famer Ted Lindsay for a special pre-game ceremony at old Olympia Stadium in Detroit. Lynch was a two-time president of the Detroit Sports Broadcasters Association and worked every game played at the old Olympia. He also broadcast on stations WWJ-AM, WJR-AM, CKLW-AM and WXYZ-TV, and was elected to the Michigan Sports Hall of Fame in 1994. Lindsay, meanwhile, would later work for NBC-TV as a hockey color analyst (next to Tim Ryan) in the 1970s.(Collection of Budd Lynch.)

Alex Delvecchio (left) tried his hand at broadcasting when he partnered with Budd Lynch for ON Television Sports games involving the Red Wings. This subscription TV outlet used the signal of WXON-TV, Channel 20, in the evenings when games were shown. Delvecchio was an all-star center for Detroit and later became the club's coach and general manager. (Collection of Budd Lynch.)

Three legends of hockey gathered when Foster Hewitt (left), the long-timer voice of the Toronto Maple Leafs and CBC-TV's *Hockey Night in Canada*, Toe Blake, the great Montreal Canadiens coach, and Budd Lynch, the Red Wings longtime radio/TV voice, are shown in this undated photo. (Collection of Budd Lynch.)

Ron Cantera (left) and Sam Nover handled intermission updates and interviews for Red Wings telecasts in the late 1960s and early 1970s on WKBD-TV, Channel 50. Cantera is seen here interviewing statistician Morris Moorawnick at Olympia Stadium. (Collection of Robert L. Wimmer.)

Red Wings broadcasters Sid Abel (center) and Bruce Martyn (standing) interview NHL referee Frank Udvari between periods in their booth high in the balcony in Olympia Stadium. Martyn was elected to the Hockey Hall of Fame in 1991 and the Michigan Sports Hall of Fame in 1996. The Sault Ste. Marie, Michigan native started his career at WSOO-AM in his hometown in 1950. He worked alongside legendary Budd Lynch on Red Wings radio and television broadcasts before succeeding Lynch in 1975. Martyn retired following the 1994–95 National Hockey League season. (Collection of Robert L. Wimmer.)

Detroit Red Wings mascot "Winger" gets a squeeze from team broadcaster Budd Lynch (right) and Gordie Howe. Budd refers to himself as the "One-Armed Bandit" and revels in delighting people with jokes based on his condition. While with the Canadian Army in 1944, near the village of Caen, France, Budd was wounded and lost his right arm. Typical of his character, Budd was undaunted by this handicap and, over the years, he's made light of his situation. He annually hosts a golf tournament with monies raised turned over to needy southeastern Michigan charities. (Collection of Budd Lynch.)

This gathering of current and former Red Wings broadcasters in the 1990s featured, from left to right, Paul Woods, Bruce Martyn, Sid Abel, Budd Lynch, and Alex Delvecchio. Woods succeeded Paul Chapman, who had earlier succeeded Abel as Martyn's radio sidekick on station WJR-AM. (Collection of Budd Lynch.)

At the franchise's 70th Anniversary bash, Red Wings broadcaster Bruce Martyn (left) was greeted by former player Stu Evans (center) and former broadcaster Budd Lynch. Martyn and Lynch worked together on radio and television, alternating two periods on one media and one on the other. Martyn has homes in Michigan and Florida and makes occasional visits to Red Wings games. (From Mark A. Hicks.)

Bruce Martyn (left) is toasted by legendary Montreal Canadiens broadcaster Danny Gallivan (center) at a Hockey Hall of Fame reception in 1992 in Toronto. Budd Lynch is also on hand. All three men are enshrined in the Hall of Fame as Foster Hewitt Award winners—Gallivan in 1984, Lynch in 1985, and Martyn in 1991. (Collection of Budd Lynch.)

Dave Strader (left), Ray Lane (center), and Mickey Redmond formed the Red Wings telecast team on WKBD-TV, UPN-50, in 1992. Missing from the photo is longtime producer Toby Cunningham. Strader toiled with the minor league Adirondack Red Wings before earning a shot in the NHL with the parent Detroit team, succeeding Bruce Martyn, who stayed on to do radio play-by-play. Strader now works for ESPN doing NHL telecasts. Redmond completed his 17th season with the team in 2003. (Collection of Ray Lane.)

Mike Emrick of St. Clair, Michigan has become one of hockey's most noted voices. Shown here while working for CBS Sports on NHL Game of the Week telecasts, Emrick is the dean of American hockey broadcasters, with 30-plus seasons under his belt. He's worked for ABC, ESPN, Fox, and the New Jersey Devils television network, and broadcasted his 2,000th professional hockey game on April 3, 1999 (New Jersey at Pittsburgh). Mike was the radio/TV and public relations director of the Port Huron (Michigan) Flags from 1973 to 1977. (From CBS Sports.)

Harry Neale is a former coach of the Detroit Red Wings and now one of hockey's brightest TV commentators, heard on CBC-TV's *Hockey Night in Canada* telecasts nationally. Teamed with Bob Cole, Neale possesses a great knowledge of the league and is used by CBC on all its major events such as the NHL All-Star Game, Stanley Cup playoffs, and Stanley Cup Finals. While he couldn't succeed in Detroit as a coach, Neale has made it big as an excellent commentator. (From Detroit Red Wings.)

Detroit Red Wings radio voice Ken Kal presented the Detroit Sports Broadcasters Association Detroit Tigers Rookie of the Year Award to Matt Anderson at Tiger Stadium in 1999. Kal succeeded Bruce Martyn as lead broadcaster of the Red Wings in 1995 and previously called University of Michigan hockey games on WTKA-AM. He's a graduate of Wayne State University. Ken calls his job with the Red Wings "a dream come true." He was DSBA president from 1999 to 2001. (From Ben Eriksson.)

Paul Woods (left) has racked up 16 seasons as the Detroit Red Wings radio network's color analyst after a playing career that included seven years in a Detroit uniform (1977-84). He finished his career with 72 goals and 124 assists in 501 NHL regular-season games. Woods tries to trip up play-by-play man Ken Kal (right) on hockey trivia during each broadcast. (From Detroit Red Wings.)

Mickey Redmond (left) once scored 50 goals as a player and Ken Daniels (right) once appeared in a made-for-TV movie called *On Thin Ice—The Tai Babilonia Story*. Together, they form the Red Wings telecast team on Fox Sports Net and WKBD, UPN-50. Both once worked for CBC's *Hockey Night in Canada*. Redmond (17 seasons) and Daniels (6 seasons) have proven to be popular and successful in hockey-crazed Detroit. As lead announcer, Daniels has the experience of having worked in hockey hotbeds Toronto and Detroit while Redmond played for Montreal and Detroit. When Daniels was not covering hockey, he did auto racing, golf, the Olympics and sports-talk in Canada. Redmond owns his own travel business. (From Fox Sports Net.)

Art Regner (third from left) is one Detroit's most fervent hockey broadcasters, handling the between-periods, pre-game and post-game shows on flagship station WXYT-AM for Red Wings games. A native Detroiter, Regner developed a passion for hockey at an early age, and has followed the Red Wings for a lifetime. His "GO WINGS" shouts are now staples with callers and listeners of his shows. Regner was part of the original crew of broadcasters and producers that started WDFN-AM but was wooed to WXYT as the station landed the Red Wings radio broadcast rights. (From WXYT-AM.)

The "Cold War" wasn't just fought between the United States and the Soviet Union. Fox Sports Net covered the record-setting college hockey game featuring the Michigan State Spartans and University of Michigan Wolverines in November of 2002 at cavernous Spartan Stadium. As seen here, from left to right, Shireen Saski, Matt Shepard, John Keating, Billy Jaffe, and Trevor Thompson reported from East Lansing, on an ice surface that was placed over the football field. (From Fox Sports Net.)

Barry Melrose spent a brief time in a Red Wings uniform and once coached the Wings' top farm club in Adirondack, NY. He also coached in the NHL with the Los Angeles Kings. Melrose has really made a name for himself as the top hockey studio analyst for ABC-TV and ESPN, teaming frequently with John Saunders. (From ABC/ESPN.)

No chapter on hockey in Detroit is complete without mentioning Sonny Eliot (left), shown here with Ted Lindsay, Hockey Hall of Fame player. Eliott was given a *one dollar* contract by the Red Wings to play goal in practice during the 1950s and 60s. He was the full-time weather person on WWJ-TV and WWJ-AM while also covering Red Wings and Tigers games. (From Ben Eriksson.)

In this 1999 photo, Red Wings broadcaster Ken Kal (far right) emcees a "Red Wings Alumni Roundtable" meeting of the Detroit Sports Broadcasters Association in Farmington Hills with former Detroit skaters (left to right) Ted Lindsay, Johnny Wilson, and Bill Gadsby. All three also coached or managed the Red Wings at one time. (From Bob Benko.)

FOUR
Basketball

Little did Paul Carey (left) know when he had Detroit Pistons General Manager Nick Kerbawy on his High School Basketball Scoreboard Show on WJR-AM that one day he would be the play-by-play voice of the Pistons. In this late-1950s photo, Carey talks hoops with Kerbawy while thanking him for the sponsorship of the show, a long tradition at the station called "The Great Voice of the Great Lakes." High school basketball coaches from throughout southeastern Michigan and points beyond in the state would phone scores and highlights to Carey for late Friday night broadcast to thousands of listeners. (Collection of Paul Carey.)

Milt Hopwood (right) called Detroit Pistons games for several seasons. In this 1969 photograph, Hopwood works alongside Paul Carey (center) and engineer Howard Stitzel (left) courtside at Cobo Arena in downtown Detroit. Basketball affords most team broadcasters a front-row seat to the action. Tom Hemingway was also a top Pistons broadcaster. (From James D. McCarthy.)

John Fountain has been voice of Eastern Michigan University basketball for 38 seasons and is in his 54th year of broadcasting. The former president of the Detroit Sports Broadcasters Association has distinguished himself for his basketball—as well as football—play-by-play at EMU in Ypsilanti, Michigan, as heard on WEMU-FM. John broadcast his first event (a football game) from Memorial Stadium in Port Huron, Michigan in 1950. In the mid-1950s he reported sports for Armed Forces Radio from a base in Tripoli, Libya. It was there that he did a re-creation of the seven-game World Series between the New York Yankees and Brooklyn Dodgers on a ten-second delay. (Collection of John Fountain.)

As well known for his basketball broadcasts as he is for calling football action in the state of Michigan, George Blaha is a name and voice most familiar. Named 1998 Michigan Sports Broadcaster of the Year, Blaha has completed 26 years of Pistons basketball on flagship WDFN-AM, and 25 years as the voice of Michigan State Spartans football on WWJ-AM & WXYT-AM. He's shown here interviewing NBA Commissioner David Stern for his Pistons pre-game show. (Collection of George Blaha.)

Dick Vitale started his broadcast career in Detroit. As head coach of the University of Detroit Titans, Vitale attracted national attention with his enthusiasm and passion for his team and the game of basketball. His success with the college Titans earned him a chance to coach the Detroit Pistons of the NBA. Shown here in his Pistons duds, Vitale was invited by WXYZ-AM to host a weekend *Sportstalk* show on the station. While he failed to deliver a winner to the Pistons, Vitale was a success on radio. He eventually landed big-time broadcasting gigs with ESPN and ABC. The rest, as they say, is history. (From Detroit Pistons.)

The Detroit Pistons marketing department produced a vinyl (lightweight) record in 1979 of Dick Vitale extolling the virtues of the Pistons. Shown here is the cover jacket with Vitale smiling in the center. The player wearing No. 16 is Bob Lanier, and wearing No. 25 in John Long. (From Detroit Pistons.)

When not broadcasting a game, George Blaha often travels to Detroit Pistons games aboard "Roundball One," the team's airplane. In this candid photograph, we see a very tired Pistons broadcaster getting some rest. We're just surprised to see he didn't pull the shades down! (Collection of George Blaha.)

One of George Blaha's most colorful color commentators (and he's had a lot of them) was John "Crash" Mengelt (left). The duo is shown at one of basketball's old shrines, the Boston Garden, prior to a WKBD-TV broadcast. Mengelt, a guard from Auburn University, played four exciting seasons in Detroit, averaging 10.3 points per game. He was acquired in a trade by the Pistons with Kansas City-Omaha in 1972, and left the Pistons following the 1975-76 season. (Collection of George Blaha.)

One of the greatest players in basketball history meets one of the top broadcasters. Julius "Dr. J" Erving (left) greets George Blaha prior to a Detroit Pistons-Philadelphia 76'ers game. "Blahaisms" include "He fills it up," "He blew the bunny," "Write this one down, Mrs. Brown," and "Off the high glass." (Collection of George Blaha.)

George Blaha (left) gives an award to Vinnie Johnson of the Pistons. George gave Johnson his "007" nickname because the Pistons' guard hit a game-winning jumper in the deciding game of the 1990 NBA Finals at Portland with 0.07 remaining on the game clock. (Collection of George Blaha.)

Dick Motta (left) is one of the winningest coaches in NBA history and so Pistons broadcaster George Blaha (right) nicknamed him "Dick Motta, Mr. 808" when he worked WKBD-TV games, signifying the 808 lifetime victories at that juncture. NBA Commissioner David Stern is the man in the middle. (From Michael P. Callahan.)

John "Spider" Salley won two NBA Championship rings as a forward for the Detroit Pistons. He parlayed that success into a broadcasting career—but not here in Detroit. Salley had a short stint on the NBA on NBC, serving as a studio analyst. He's been seen on Fox Sports shows and other Hollywood programs as a guest. (From Detroit Pistons.)

Doug Collins coached Michael Jordan and he coached the Pistons yet failed to win a championship with either. So what do former coaches do in many cases? They go to the broadcast booth, as in the case of Dick Vitale, Dick Motta, Kevin Loughery, Jack Ramsey, Tommy Heinsohn, Matt Goukas, Mike Fratello, Hubie Brown, P.J. Carlisimo, and Chuck Daly. Collins moved up the ladder fast at NBC-TV when he joined the No. 1 broadcast team with Bob Costas and, later, Marv Albert. Collins was insightful in delivering his basketball analysis. (From Detroit Pistons.)

Scott Hastings played for the Detroit Pistons just two seasons as a backup forward/center and was on the 1989 World Champions. He's made a smooth transition to basketball commentator, sideline reporter, and talk show host. He provided insight on Nuggets home-game telecasts for the fourth straight season in 2003. An 11-year veteran of the NBA, Hastings also played for Denver from 1991- 1993. A four-year letterman at Arkansas, Hastings left as the school's second all-time leading scorer (behind Sidney Moncrief) with 1,779 career points. He's one of the most well-known media personalities in the Denver market, and co-hosts a daily sports talk radio show on KOA-AM. In addition, he's in his sixth season as the color analyst for the radio broadcasts of the Denver Broncos. It is believed that he is the only person in existence to possess two Super Bowl rings and an NBA Championship ring.

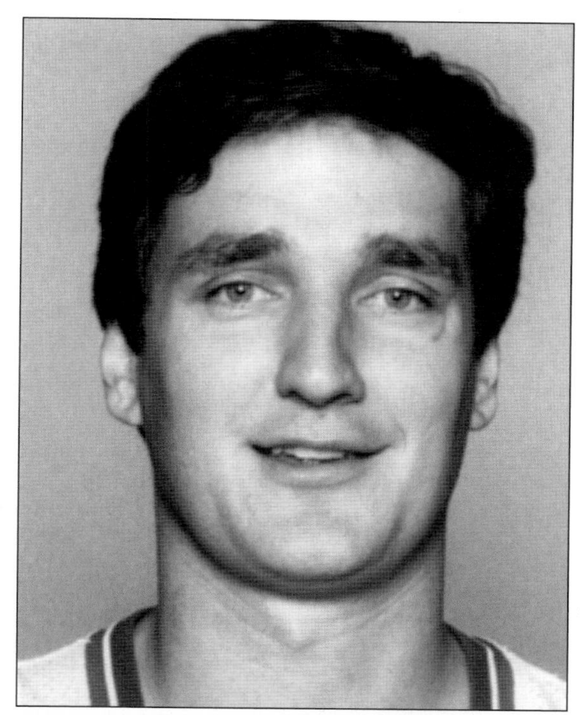

Fred McLeod (left) and Greg Kelser form the Detroit Pistons broadcast team for games shown on Fox Sports Net. McLeod completed his 20th season in 2003 while Kelser notched his 10th season on the cablecasts. McLeod also is sports anchor/reporter on WDIV-TV while Kelser works ESPN regional basketball contests plus contributes to FSN pre-game shows. McLeod's other TV credits include the Detroit Tigers, Oakland Athletics, Cleveland Indians, and Cleveland Cavaliers. (From Dan Graschuck.)

Bill Laimbeer (left) completed his first season as color analyst on Detroit Pistons games with George Blaha (right). Seen here interviewing Pistons guard Richard Hamilton, Laimbeer also is head coach of the Detroit Shock of the WNBA. "Bad Boy" Bill won two NBA titles while a Pistons player, in 1989 and 1990. (From Dan Graschuck.)

Rich Mahorn (left) has one year under his belt as a game analyst with the Detroit Pistons. Another former "Bad Boy" of the championship Pistons teams, Mahorn is seen interviewing Pistons center Mehmet "Memo" Okur with play-by-play man George Blaha, following a game at The Palace of Auburn Hills. (From Dan Graschuck.)

All-time NBA and Pistons great Dave Bing (far right) worked a couple seasons alongside George Blaha (left) on WKBD-TV. The two were reunited in Auburn Hills to honor Earl Lloyd on his 2003 induction into the Basketball Hall of Fame. Next to Lloyd are Bob Lanier (tallest person) and Will Robinson. (From Dan Graschuck.)

Will Tieman (left) and Gus Ganakas form the Michigan State Spartans radio broadcast team, as heard on WWJ-AM in Detroit. Tieman has years of Big Ten Conference experience as a play-by-play broadcaster, host, and producer of basketball and football games. Ganakas is a former Spartans basketball coach, preceding Jud Heathcote. (From Margaret Hehr.)

Notre Dame and Detroit Pistons standout Kelly Tripucka has made a fine career for himself as a pro basketball color analyst. He teamed with George Blaha in Detroit on WKBD-TV (Channel 50) before moving home to New Jersey to do color commentary on Nets games for YES Network, with Ian Eagle. Both Kelly and his father Frank are members of the National Polish-American Sports Hall of Fame. (From Detroit Pistons.)

Tim McCormick (left) played for the University of Michigan Wolverines basketball teams as well as several NBA teams and now enjoys a career in player relations and the broadcast media. He's shown working a state high school basketball game with Ray Bentley (center) for Fox Sports Net. (From Fox Sports Net.)

FIVE
Sports Anchors, Reporters, and Talk Show Hosts

This meeting of the Detroit Sports Broadcasters Association in 1950 was held at the Old Colony Club in downtown Detroit. Pictured, from left to right, are Fred Huber, Jack Teahen, Al Nagler, George Wilson, Jack Adams, unidentified, Paul Williams, Paul Pentecost, Don Hansen, unidentified, Bela de Tuscan, and Joe Gentile. The DSBA was founded by Ty Tyson and formed as a result of perceived "second-class treatment" in the Briggs Stadium press box. Their main beef was with the Baseball Writers Association of America (BBWAA). (Paul Pentecost collection.)

There weren't many sports broadcasters in Michigan with a greater presence than "Rapid" Bob Reynolds. He anchored the popular 6:15 and, later in his career, 11:15 nightly sports roundups on powerful WJR-AM. Reynolds reports from the Michigan Open Golf Tournament in 1958 in this photo. Seen here, from left to right, are Ben Lula, Reynolds, Warren Orlich, and Paul Carey. (From Bill Gallagher).

Bob Reynolds was a two-time president of the Detroit Sports Broadcasters Association and loved the game of golf. In this undated photo, he's seen interviewing PGA golf professional Bruce Crampton of Australia at a Michigan event, for WJR-AM. Paul Carey is in the sunglasses to the right of Reynolds. (Paul Carey collection.)

The WJR mobile studio was frequently found at area sports events in the 1950s and 1960s. Here's Paul Carey in front of the unit in 1958 at the Michigan Open. (From Bill Gallagher.)

Ray Lane, shown here with Paul Carey in the jubilant 1968 Detroit Tigers locker room during a live WJBK-TV feed, has been a part of Detroit and Michigan sports broadcasting for five decades, most recently as host of Red Wings telecasts on UPN-50 television and occasional announcer on Tigers telecasts on the same station. During his storied career, Lane has called Tigers, Red Wings, Pistons, and Lions games, as well as Michigan State and University of Michigan football and basketball. Ray also has done Big Ten football and basketball and Cincinnati Reds baseball. He was a two-time president of the Detroit Sports Broadcasters Association and was inducted into the Michigan Sports Hall of Fame in 1997. (Red Ruser photo.)

IF IT'S *SPORTS* NEWS YOU'RE AFTER...

If you live in the Detroit area, you are lucky. Particularly if you happen to be a television-sports fan. Why? Because Detroit not only is a hot-spot of major league activity, it also happens to be the home of two top-notch Television Sportscasters: Van Patrick and Ray Lane.

Both Ray and Van are heavily involved in broadcasting and telecasting major league activities. They both regularly do feeds to the various networks. And whether it's interviews of famous sports people, color, play-by-play (or you name it), they're busy around the calendar covering such events as Lions Football, Tiger Baseball, Big Ten Basketball, Major Golfing events, and so on.

But next to being major league Sportscasters, the important thing is that they're both local citizens. That's where you (and us, too) are so lucky. Van Patrick and Ray Lane are both available on WJBK-TV, to bring you daily sports news and information in the very best way.

On the 6 O'Clock Report each weekday, Ray Lane offers you a fast-moving package of results and stories. On the 11 O'Clock Report, Big Van brings all the "inside scoop" because he gets more exclusives than any other Sportscaster in Detroit.

And man, Van and Ray are great! That's why the networks want them for so many things. Best of all, you can hear them all week—*free!*

...THE BEST WAY TO GET IT IS ON *WJBK-TV2*

Wayne Walker (upper left) and Ray Lane, Van Patrick and Lane, and Lions player Joe Schmidt and Patrick are featured in this late 1960s WJBK-TV sports promotion insert (Ray Lane collection.)

Olympia Stadium (since razed) was the setting for this 1971 Detroit Sports Broadcasters Association charity basketball benefit against the Harlem Globetrotters. Standing, from left to right, are Bob Kouel, Paul Carey (WJR-AM), Don Lessnau, Dave Diles (WXYZ-TV/AM & ABC Sports), Larry Adderley (WXYZ-TV & WCZY-FM), Warren Pierce (WJR-AM, WYUR-AM & WKBD-TV), Bob Kiess, and Coach Ed Barbour. Kneeling, from left to right, are Ron Knight (WXYZ-AM), Tom Ryan (CKLW-AM & WOMC-FM), and Al Ackerman (WWJ-TV & WDIV-TV). Only Pierce and Ryan can still be heard regularly on southeastern Michigan radio stations. (Tom Ryan collection.)

Don Kremer joined the Detroit Tigers broadcast team in 1976 while on the staff of WWJ-TV. He began his Detroit sports broadcasting career in 1960 and his credits include Detroit Lions football, college football, bowling shows (notably *Beat The Champ* with C.A. "Buck" Walby) and the Tigers. Kremer would later become public relations director of the Lions (From Detroit Tigers.)

Ray Lane, part of WJBK-TV's "Five Star Sports" staff, interviews Earvin "Magic" Johnson of the Michigan State Spartans in this late-1970s photograph. (Ray Lane collection.)

(*above*)Dave Diles was a mainstay of Detroit radio and television much of the 1960s, 70s, and 80s before transitioning to ABC Sports assignments. His *Dial Dave Diles* radio talk show was one of the first of its kind, heard on WXYZ-AM. He was sports anchor/reporter on WXYZ-TV. Diles worked the ABC college football scoreboard show, *Pro Bowlers Tour*, *Wide World of Sports*, and other network events. As an author, he wrote *Nobody's Perfect* with Denny McLain of the Detroit Tigers. Diles steered the Detroit Sports Broadcasters Association from 1964 to 1966 and is seen here interviewing a group of Red Wings alumni. (Robert L. Wimmer collection.)

(*left*) Bill Fleming was a sportscasting regular in Detroit long before he advanced his career to ABC Sports. Shown here in his ABC blazer, Fleming was a fixture on Detroit-area stations WWJ-AM, WXYZ-TV, and WXYZ-AM, with insightful sports reports and commentary. The likable Fleming was president of the Detroit Sports Broadcasters Association from 1957-58. He worked college football, golf, Olympic Games and *Wide World of Sports* assignments for ABC. (From ABC Sports.)

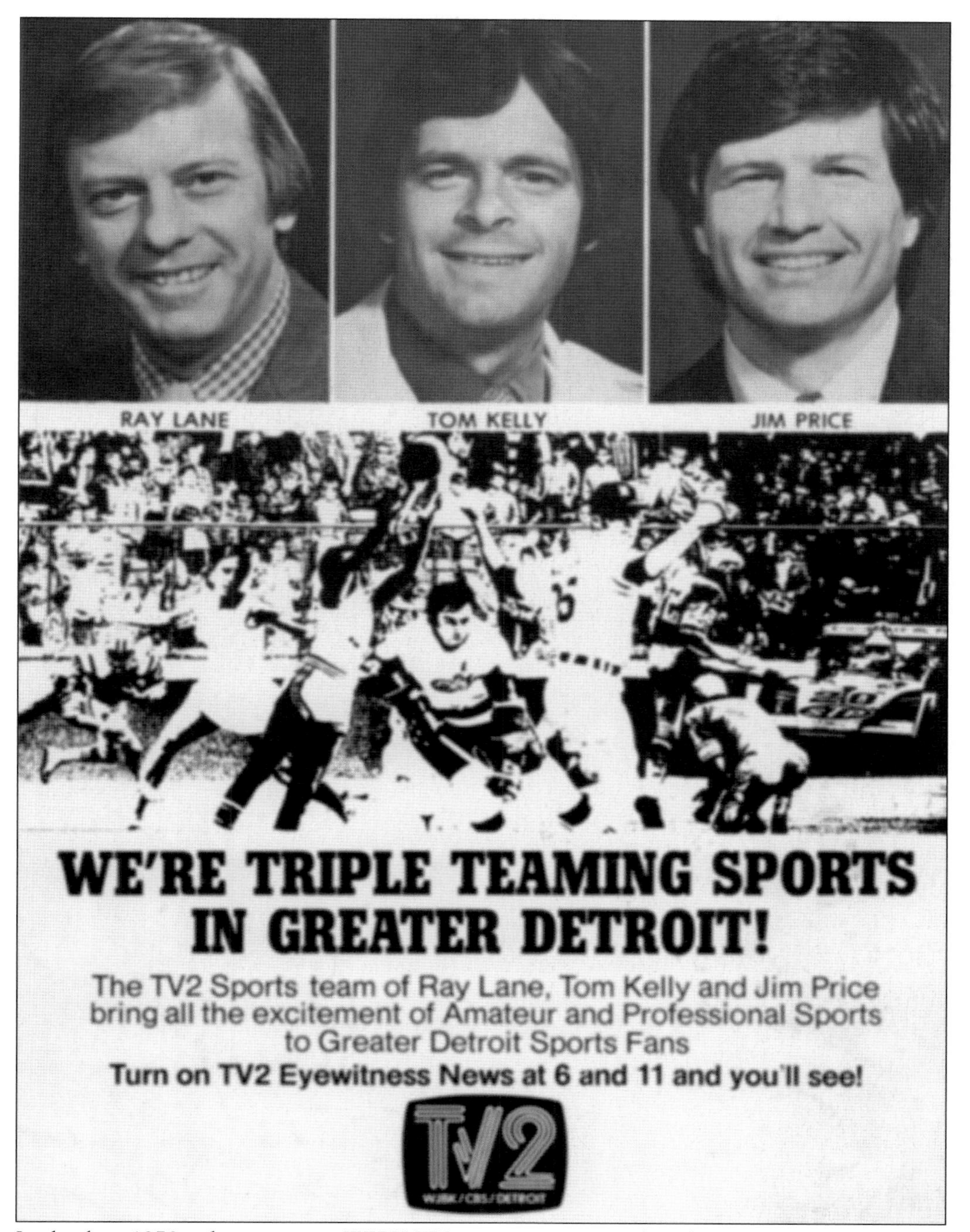

In this late-1970s advertisement, WJBK-TV trumpets its sports broadcasting crew of Ray Lane, Tom Kelly and Jim Price. (Ray Lane collection).

In 1977, WWJ-TV (Channel 4) was so proud of its affiliation with the Detroit Tigers that it commissioned an advertising agency to design this insert into the Tigers program, featuring broadcasters (left to right) Larry Osterman, George Kell, Joe Pellegrino, and Al Kaline. (From Detroit Tigers.)

THE HEART OF THE ORDER

Follow our lineup. And you've got a free seat for 50 of this season's exciting Tiger games on TV. Al Kaline and George Kell bring you every play-by-play. They're part of our Sports Corps.

We bring you all the sports Detroit plays. Sports Corps reporters Mike Barry, Gail Granik, Don Shane, and Jim Brandstatter tell the sports story the way you want to hear it. We're for people who love sports. Pro or amateur, college or minor league. Seven nights a week on News 4 Detroit.

Sports Corps reporters:
Al Kaline George Kell
Mike Barry Don Shane
Gail Granik Jim Brandstatter

WE'RE 4 SPORTS
WDIV DETROIT

WWJ changed its call letters to WDIV by the time this 1979 ad was placed inside the Detroit Tigers program. Four new faces—Mike Barry, Don Shane, Gail Granik and Jim Brandstatter—were on-air announcers for the station that still featured the Tigers play-by-play team of George Kell and Al Kaline. (From Detroit Tigers.)

Joseph Priestly "J.P." McCarthy (left) was technically never a sports broadcaster in Detroit, but you'd hardly know it. On his No. 1-rated WJR-AM morning program he talked sports every hour on the hour it seemed. McCarthy is shown here emceeing a late 1970s March of Dimes awards banquet at the Raleigh House, honoring Mark "The Bird" Fidrych (at podium). (From Bill Axtell.)

Mark "Doc" Andrews (left) enjoys a pre-game chat with Barry Smades (center) of the Oakland Press, and Olympia Stadium press box manager Hugo Costillo in the late 1970s. Andrews has worked for numerous radio and TV stations in metro Detroit including his current association with "Purtan's People" on WOMC-FM. The Wayne State University graduate has done public address announcing for all four of Detroit's big league teams. (Robert L. Wimmer collection.)

Don Daly (right) was seen at major events in Detroit and Windsor, Canada, for two decades, as sports anchor/reporter for CKLW-TV, Channel 9. Daly is seen here interviewing Red Wings General Manager Alex Delvecchio at Olympia Stadium in Detroit (date unknown). (Robert L. Wimmer collection.)

Ron Cameron (left) was one of the most controversial and bombastic sportscasters in Detroit's history. His resume include stints on WCAR-AM, WEXL-AM, WPON-AM, WHND-AM, WMZK-AM, WXYZ-AM, WXON-TV, and WGPR-TV. His guests were top-notch and his comments always insightful. He's seen here interviewing Red Wings defenseman Terry Harper in the late 1970s. (Robert L. Wimmer collection.)

Known as one of Detroit's brashest sports broadcasters, Al Ackerman of WWJ-TV (later WDIV-TV) is shown here at a 1980s Detroit Tigers Winter Party at the Detroit Athletic Club with Jim Campbell, general manager of the Detroit Tigers. Ackerman wasn't afraid to grill Campbell, Lions GM Russ Thomas or any other sports figure. Ackerman also worked for WWJ-AM and WXYZ-TV and is now retired to Sanibel Island on Florida's gulf coast. (From Charles Jackson.)

Another colorful radio sports personality in Detroit was Vince Doyle (left) of WWJ-AM. The Irishman was witty and hard-working. Doyle was president of the Detroit Sports Broadcasters Association from 1972 to 1974. His daughter, Anne Doyle, was the area's first female sportscaster, on WJBK-TV. Vince is shown here with Dick Beddoes of CBC-TV's *Hockey Night in Canada*, and Red Wings publicist Al Coates. Doyle also worked Michigan Stags hockey and University of Michigan football broadcasts. (Robert L. Wimmer collection.)

Jim Forney loved his soccer as well as his broadcast days in Detroit. He earned all-state soccer honors in Pennsylvania and won a soccer championship in his college division in 1961. It's no wonder he became voice of the Detroit Express soccer team, and overseer of the club's broadcast functions. Before landing in Detroit, he worked in Pittsburgh and was the voice of the Penguins. Forney anchored sports at WWJ-TV, Channel 4, and later at WJR-AM. (From Glenn L. Hibbert.)

She was known as Gail Granik in Detroit (on WDIV-TV) and later became Gayle Gardner when she joined the staff of NBC Sports. Gardner is shown in this NBC/Cablevision's Olympics TripleCast team photo with, clockwise from left to right, Bruce Jenner, Frank Shorter, Don Criqui, Gardner, Kathleen Sullivan, Julianne McNamara, and Ahmad Rashad. (From NBC Sports.)

Chris McClure (left) has been synonymous with boxing and motor sports particularly during his on-air days on Detroit television and radio. He's seen here interviewing manager/trainer Emanuel Steward and Tyrone Trice (date unknown) on Pro Am Sports System (PASS). McClure called Detroit Grand Prix open-wheel races, hydroplane races, and many other events on PASS, WCXI-AM, ESPN, ESPN2, and The Speed Channel. McClure was president of the Detroit Sports Broadcasters Association from 1979 to 1981. (Lindy Lindell collection.)

Handling sports updates on WWJ-AM in the early 1980s was Fred Manfra. This personable reporter and anchor also called University of Michigan Wolverines football with Vince Doyle on the station. Manfra would later work for ABC Radio Sports, anchoring weekend scoreboard shows, horse racing's Triple Crown events, and the Winter and Summer Olympic Games. (From ABC Sports.)

The Detroit Sports Broadcasters Association presented an "Outstanding Service Award" to past president Joe Gentile in 1980 at the Detroit Press Club. On hand, from left to right, are Barry Smades, Bernie Reilly, Kurt Schneider, Gentile, Leo Martin, Ray Lane, and Bob Kiess. (Bob Kiess collection.)

Jack Riggs has called thoroughbred and harness racing in southeastern Michigan, and Windsor, Canada, for more than 40 years. He was heard on WWJ-AM for over 12 years with his *Race Track Report* show, heard five nights a week. Riggs produced and announced a one-hour, five nights a week TV show on Pro Am Sports System (PASS) that was also seen on Comcast cable. In the 1950s, Jack worked with Bob Maxwell on the first TV show to air from a Michigan track when they broadcast the big race once a week from Northville Downs on WWJ-TV. (Jack Riggs collection.)

Norm Plummer was a popular sports anchor and talk show host on WWJ-AM in the 1970s. The colorful Plummer hosted a call-in show week nights when those formats were in their infancy. He later moved to Toledo and became a much-listened-to disc jockey where he went by the name of Bob Kelly. In 1974, Plummer was the radio voice for the Michigan Stags of the World Hockey Association. Plummer is responsible for putting Ron Cameron on the airwaves during their "Ask the Umpires" segment on WWJ.

Eli Zaret (right) began his broadcast career in Detroit in 1974. Shown here after a hoops challenge with partner Fred McLeod in a WJBK-TV promotion, Zaret's distinctive scratchy voice has been heard on WJBK, WDIV-TV, WABX-FM, WJZZ-FM, WRIF-FM, WABC-TV (New York City), and most recently, co-hosting *The Locker Room* show on WXYT-AM with former Detroit Lions quarterback Gary Danielson. Zaret has won numerous Associated Press, United Press International and local Emmys for his sports reporting (From WJBK-TV.)

Fred Heumann began his commercial broadcast career in 1980 at WJIM-TV in Flint. The Central Michigan University graduate was hired in 1987 to work weekends and file sports reports for WJBK-TV. He spent seven years at Channel 2 before jumping to WDIV-TV in 1984. He added WJR-AM to his duties in the early 1990s, working University of Michigan football and basketball games and co-hosting "SportsWrap" with Steve Courtney. Now Heumann is with WLNS-TV in Lansing and part-time with WXYZ-TV, Channel 7 in Detroit. (From WJBK-TV.)

Kurt Schneider (the one without the mask!) is one of Detroit's most durable sports broadcasters. He's seen here with professional wrestler White Lightning prior to a 1991 match that Schneider organized. Schneider operates Sports Phone (Phone Programs) and MCW Communications and has been Wayne State University's hockey coach and football broadcaster. He's also worked at WJR-AM, WPON-AM and WBRB-AM. Kurt is a two-time president of the Detroit Sports Broadcasters Association. (Kurt Schneider collection.)

Sports talk and sports broadcasting run in Rich Kincaide's blood. He worked at WJR-AM for several years in the 1980s and 90s before taking play-by-play assignments with minor league hockey teams in Grand Rapids and Toledo. Kincaide was pregame, intermission, and post-game host for several seasons of Red Wings hockey on WJR. He was president of the Detroit Sports Broadcasters Association (1995–96).

Fred Hickman (left) spent several years at WDIV-TV (Channel 4) in Detroit before overcoming an admitted substance abuse problem and moving south to Atlanta where he eventually settled in as host of TBS' Goodwill Games telecasts from Moscow in 1994 with Nick Charles. Hickman and Charles also hosted the popular nightly sports show on CNN for many years. (From Turner Sports.)

Andrea Joyce—shown here with Detroit radio personality Tom Ryan—never did much sports while on Detroit television but has become an excellent network sideline and feature reporter. Joyce excelled in the news department at NBC affiliate WDIV-TV for years before getting her big break and moving on to CBS Sports. She contributes to NFL football and NCAA college football and basketball telecasts on the network. (Tom Ryan collection.)

Bob Page (right) was part of the sports landscape in Detroit for over two decades. Page worked with several media outlets and earned a reputation as a hard-nosed sports reporter. He was at WJZZ-FM, WRIF-FM, WJR-AM, WXYZ-TV, and as co-host of "Sports View Today" with Ron Cameron (left) on several cable stations and WXON-TV (Channel 20). Page replaced a legend when he succeeded Howard Cosell on ABC Radio and his "Speaking of Sports" commentaries. Page spent considerable time at Madison Square Garden Network and Fox Sports.

Bowling shows have been popular in metro Detroit over the years. Tom Ryan (center with microphone) hosted the *Michigan Bowling Challenge* on WKBD-TV, Channel 50, in 1992. Others who have hosted bowling shows include Fred Wolf, Don Kremer, Buck Walby, Bob Allison and Tom Mazawey (Pax Network). Ryan is the longtime public address announcer of the University of Detroit basketball Titans, and former PA voice of the Detroit Lions and Pistons. He has also hosted an afternoon music show on WOMC-FM in recent years. (Tom Ryan collection.)

Tom Mazawey is executive producer of sports at WJR-AM. His responsibilities include daily sports updates, the *Sportswrap* show, college scoreboard shows and reporting duties at the station. He's a New York Yankees and New Jersey Devils fan yet that won't stop "Maz" from enjoying Tigers and Red Wings games too! Tom knew he wanted to go into radio when he grew up listening to WABC in New York City as a youngster. (From WJR-AM.)

One of Detroit's most popular sports broadcasters since 1986 is Bernie Smilovitz of WDIV-TV, Channel 4, shown here interviewing Detroit Lions Coach Steve Mariucci. The Brooklyn, New York native hosts his "Weekend at Bernie's," "Bernie's Blooper," and "We've got highlights" segments on the station's weekday evening sportscasts. With the exception of two brief years in New York City, Smilovitz has been a Channel 4 mainstay. His Detroit Tigers pregame show segments with Manager Sparky Anderson were often entertaining. Previous to Channel 4, Bernie worked at Washington, DC station WTTG-TV. (From Dan Graschuck.)

Don Shane has called Detroit home much of his brilliant broadcasting career. He's tireless in pursuit of a sports story and most often takes his evening shows for WXYZ-TV, Channel 7, to venues such as Comerica Park, Ford Field, Joe Louis Arena, and The Palace of Auburn Hills. Shane (lower right) is shown interviewing golf great Tiger Woods at the Buick Open near Flint, Michigan. He joined Channel 7 as sports director in 1986 but had previous experience in the market with Channel 4. Shane also worked at WBZ-TV in Boston and WMAQ-TV in Chicago. (From Al Abrams.)

Dan Miller is the accomplished sports director at WJBK-TV, Channel 2, in Detroit and can also be seen on Fox Sports' coverage of National Football League regional games. Miller hosts the Sunday night *Sports Works* program as well as pre-game and post-game Detroit Lions shows on the station. He's also heard on a weekly pro football radio show with Craig James. Miller joined Channel 2 in 1997 after working in the Washington, DC market at two stations and hosting a daily radio talk show on WTEM-AM there. (From WJBK-TV.)

Sports has been in the blood of Dave LewAllen since his days at Central Michigan University. Like Dick Enberg and Fred Heumann, this CMU graduate made the logical move to sports broadcasting. His credits include WJR-AM, WJBK-TV, WMTG-AM (doing University of Detroit basketball), and WXYZ-TV where he's been employed since 1988. LewAllen most recently has added weekend news co-anchoring to his sports reporting duties. He was president of the Detroit Sports Broadcasters Association from 1983–85. (From WXYZ-TV.)

Ike "Mega Man" Griffin was an outside linebacker at Michigan State University and played in the NFL, USFL, and Canadian football leagues. He began his career in radio as an account executive for WJLB-AM and later became a sports reporter and talk show host at WJZZ-FM. Next came two years at all-sports WDFN-AM plus four years as a feature reporter on WXYZ-TV. Then Griffin was heard on WXYT-AM and later signed on with WKBD-TV (Channel 50) as a weekend sports anchor and reporter. He's since moved out of the Detroit market.

When it comes to pre-game shows and features during a game, there's not many better than John Keating of Fox Sports Net. Seen here during a live telecast from Comerica Park prior to a Detroit Tigers game, Keating is seen before every Tigers and Red Wings game which FSN airs. He served in a similar role as host of *Game Night* on Pro Am Sports System (PASS) in 1996-97. Keating began his broadcast career while a student at Grand Valley State University in Michigan. He spent 10 years working in Denver, concluding as studio host on Denver Nuggets and Colorado Avalanche radio networks. (From Fox Sports Net.)

He declined an offer to try out for the Dallas Cowboys to pursue a career in journalism, and it has paid off for Jay Berry of WXYZ-TV, Channel 7. For nearly two decades, Berry has seen it all in Detroit as a reporter, weekend anchor, and host of the Sunday night highlights show. The personable Berry is polished and professional in his approach to coaches, managers, and players. He was named to the sophomore All-American team while playing football for the University of Wyoming. (From WXYZ-TV.)

A familiar voice to Detroit sports is that of Greg Russell. Seen here working a state high school basketball championship game for Fox Sports Net, Russell was an original member of all-sports WDFN-AM when it debuted in 1994. He teamed with Lary Sorenson for the morning-drive sports talk show before being reassigned. Russell has most recently worked for WXON-TV (Channel 20) doing news and sports updates. (From Fox Sports Net.)

From Chicago to Detroit to Toronto, sports fans know of Chuck Swirsky. Although his tenure at WJR-AM was not as long as expected, Swirsky was much listened to, as sports director and voice of University of Michigan basketball during his tenure. The Detroit stop was sandwiched around 12 years at WGN-AM in Chicago and the last five years in Toronto as play-by-play voice of the Raptors of the NBA. His career dates back to when he was 12 years old, updating statistics at KFKF Radio in Bellevue, Washington. (From WJR-AM.)

A top-notch Michigan native to hit it big nationally is Mike Tirico of ABC and ESPN. Since joining ESPN in 1991, after four years at WTVH-TV in Syracuse, New York he has covered major sporting events like the Super Bowl, NCAA Final Four, College Football National Championship game, NBA Finals, NFL Draft, and the Daytona 500. Tirico earned an Emmy nomination in 2001 as top host/play-by-play for his work anchoring ABC's golf coverage that includes the British Open. He also hosts the NBA studio show on ABC and does NBA play-by-play on ESPN as well as college football announcing. Armen Keteyian of CBS and HBO Sports is another local guy made good! (From ABC Sports.)

Another member of the ABC team with Detroit roots is Gary Danielson (center). The former Detroit Lions quarterback is seen here co-hosting *The Locker Room* on WXYT-AM with Eli Zaret (left). They are interviewing Lions Coach Steve Mariucci at Ford Field. Danielson's star has risen fast at ABC (since 1997) where he is the No. 1 college football analyst. His previous experience includes WDIV-TV in Detroit and stations in Cleveland—where he played for the Browns. (From Dan Graschuck.)

Paul Chapman was a part of the WJR-AM sports department for many years in the 1980s and 90s and worked one season as radio color analyst for Detroit Red Wings on the station. "Chappy" was a busy sports "street" reporter and often anchored the 6:15 p.m. and 11:15 p.m. sportscasts, in addition to working with Bruce Martyn one hockey season. He arrived at WJR from Ann Arbor where he called Michigan Wolverines games. Chapman was president of the Detroit Sports Broadcasters Association from 1986-88. (From WJR-AM.)

He was colorful and off-beat yet could never quite catch on with the Detroit television public. Van Earl Wright came to Detroit's WDIV-TV from CNN in late 1993 but his stay was short-lived. The Atlanta native made signature calls of plays on CNN's *Headline Sports* in 1989 and in 1990 was named CNN's sports anchor. Wright's sports broadcasting career has stopped in South Carolina, Mississippi, Texas, Georgia, Michigan, and California—where he recently worked for Fox Sports. He also worked as a WDFN-AM (Detroit) talk show host for a while. (From WDIV-TV.)

One of the most versatile sports broadcasters in Detroit is Jim Berk. Just about every station in town has called on him to anchor or report a sports story during the last decade or so. Berk's emphatic delivery of University of Detroit basketball on WTMG-AM radio was memorable. His full-time and freelance TV stints on WDIV, WKBD, WJBK, and WXYZ make Berk our "Marathon Man" on Detroit sports broadcasting charts. He also anchored sports at many radio stations including WWJ-AM and WJR-AM.

Whether anchoring sports or hosting a talk show, Dan Gutowsky is at ease. The WWJ-AM & WXYT-AM sports announcer has experience as host of Michigan State Spartans, Detroit Lions, and Detroit Pistons wraparound programs. He also has anchored pro and college football scoreboard shows for the two stations. (From WWJ-AM.)

Tim Swore (top) and Chuck Garfien (bottom) are the last two sports directors employed by the WKBD-TV (Channel 50) programming department. Each also had the opportunity to be part of Detroit Red Wings coverage on the UPN affiliate during the recent success of the team. Swore lasted longer than Garfien, who was a victim of Channel 50's decision to no longer produce its own news programs in 2002. (From WKBD-TV.)

Dale Conquest's voice was heard for over a decade on WJR-AM and WWJ-AM during the 1980s and 90s, doing sports anchoring and reporting. Later, he lasted a short time on WYUR-AM. Conquest also called some football and basketball play-by-play on University of Michigan Wolverines games.

In 1988, Steve Garagiola (left) sat down with his father, Joe Garagiola, before Joe was to broadcast a Detroit Tigers game nationally on NBC-TV. Steve has been a mainstay in Detroit television sports and news since 1980. He worked for nine years at WXYZ-TV as sports anchor/reporter and then switched to KTSP-TV in Phoenix. Steve began his stint at WDIV-TV as a sportscaster for the station's PASS Sports network in 1994. He switched from sports to news several years ago and co-hosts Local First News weekday mornings from 5-7 a.m. on Detroit's Channel 4.

Bill Humphries (seated right) was one of Detroit's best known African-American sports anchors. As sports director of WGPR-TV (Channel 62), Humphries provided updates and hosted sports interview shows on the independent, minority-owned station. He was president of the Detroit Sports Broadcasters Association from 1978-79, and is shown here receiving an honorary lifetime membership in 1997 from DSBA President Jim Brandstatter. DSBA officer Dennis Davidson looks on.

Named the nation's No. 1 sports columnist time and again, Mitch Albom hosts the popular *Monday Sports Albom* show Monday evenings on WJR-AM for 12 years running. In addition, he's host of the *Mitch Albom Show* weekdays during afternoon-drive on the station. Albom's Monday show guests are some of the biggest names in sports, contacts he's made over years while at the *Detroit Free Press*. Mitch also is seen on ESPN's *The Sports Reporters* roundtable and wrote the best-seller, *Tuesdays with Morrie*.

Jennifer Hammond (left) of WJBK-TV (Channel 2) has become a popular sports reporter and anchor during her six-year stint at the station. Seen here broadcasting the Red Wings Stanley Cup Championship parade with Trevor Thompson of Fox Sports Net, she has strong ties to the state. Although Hammond was born in Lake Forest, Illinois, she was raised in Birmingham, Michigan, attended Seaholm High School there and received her degree at Western Michigan University. She worked at various Chicago radio stations before settling in at WDFN-AM in 1994 as a sports reporter and anchor. Nicknamed "The Hammer," she works sidelines on Fox Sports' NFL telecasts. (Jennifer Hammond collection.)

Gregg Henson has made a name for himself in Detroit radio circles. His opinionated and combative style of sports talk-show hosting has carried him to prominence at two competing all-sports stations. Henson began at WDFN-AM and brought that station success as its program director and morning-drive co-host (with Jamie Samuelson). Then came Henson's 2003 switch to all-sports WXYT-AM and subsequent midday talk-show spot with former WDFN colleague Art Regner.

Woody Woodriffe (right) of WJBK-TV (Channel 2) interviews Pro Football Hall of Famer Lem Barney at Ford Field for his station's evening sportscast. Woodriffe came to Fox-2 from WFOR-TV in Miami where he was an anchor, reporter, producer, and photographer. He is a 1984 graduate of the University of Miami. His duties at WJBK include reporting and anchoring sports. (From Dan Graschuck.)

Matt "The Diesel" Dery is sports director of all-sports WDFN-AM, afternoon updates anchor, and host of Pistons games on the team's radio network. He's also part of the Detroit Shock broadcast team and calls play-by-play for the University of Detroit Mercy Titans on WXDX-AM. The Cleveland native graduated from Syracuse in 1995. (From WDFN-AM.)

Mike Stone (left) and Bob Wojnowski host the popular *Stony & Wojo* radio sports talk show, heard during afternoon-drive on WDFN-AM-The Fan. Philadelphia native Stone worked as a sports producer at WDIV-TV and on the *Sunday Sports Albom* show on WJR-AM before joining WDFN in 1994. Their commentary, interviews, humor, and zany contests combined with timely guests make the show a hit with Detroit listeners. Stone also contributes to Channel 7's *Sunday Sports Update,* while Wojnowski is a sports columnist for *The Detroit News* and frequently guests on WJBK-TV's *SportsWorks* and Lions pre-game shows. (From WDFN-AM.)

The *Detroit Sports Report* on Fox Sports Net is a dependable daily digest of sports news. Angie Arlati anchors the program with Mickey York from studios in Seattle. She joined FSN in 1998, reporting on the Seattle Mariners and Seahawks from FSN's Northwest affiliate. Before entering the broadcast profession, Arlati was a three-time college All-American softball player at the University of Washington and later played professional baseball as an original member of the Colorado Silver Bullets. (From Fox Sports Net.)

Parker and The Man is one of Detroit's longest-running sports talk shows. Hosted by Rob Parker (below) and Mark Wilson (right), the program is heard weekday evenings on WKRK-FM in Detroit. Guests frequent the program. Wilson is the former WJBK-TV sports director and anchor and one-time WDFN-AM on-air personality. (From WKRK-FM.)

Eight presidents of the Detroit Sports Broadcasters Association gathered in November 1998 at the Novi, Michigan, Hilton hotel to mark the 50th anniversary of the association. Shown from left to right are Paul Chapman, Kurt Schneider, Gary Ratski, Jim Brandstatter, Larry Adderley, Budd Lynch, Ray Lane, and John Fountain. Other luminaries on hand for the event included

sportscasters Ernie Harwell, Fred Heumann, Mark Unger, Jim Stark, Ed Kaltz, Dave Geraci, and Scott Morganroth; weathercaster Sonny Eliot, boxer Thomas Hearns, and hockey great Ted Lindsay.

Jamie Samuelson (right) is one-half of the *Jamie and Brady Show* heard weekday mornings on WDFN-AM. Greg Brady (left) is his co-host. Samuelson is a California native and graduate of Northwestern University. He began his WDFN tenure as a sports update anchor in 1994 and became a full-time talk show host a year later. He's also the voice of Detroit Fury indoor football on the station. Greg is a huge hockey enthusiast who also started on WDFN as an update anchor. Brady also co-hosts the *Ice Time* hockey show on WDFN. (From WDFN-AM.)

Trevor Thompson (center) of Fox Sports Net is one of FSN's most durable sports reporters—seen at Lions, Tigers, Red Wings, and Pistons practices and games. Seen here interviewing Lions Coach Steve Mariucci (left) with Rob Rubick, Thompson came to FSN in Detroit following a successful stint at CTV Sportsnet in Canada. His credits also include TSN, the NBA Vancouver Grizzlies, and Headline Sports in Vancouver. (From Dan Graschuck.)

Vito LoPiccolo (left) has Detroit's longest-running motor sports show, *Motor City Motor Sports*, heard on WLQV-AM. He is shown interviewing Jack Roush of Roush Racing. LoPiccolo is a retired Detroit Police officer who has worked at MIS and been active in the Detroit Sports Broadcasters Association. (Vito LoPiccolo collection.)

A familiar voice to auto racing fans and morning-drive listeners on WWJ-AM is that of Larry Henry (left). His baritone voice is synonymous with race calls from Indianapolis, Pocono, Long Beach, Detroit, and Michigan International Speedway over two decades. Shown here interviewing NASCAR Winston Cup Series driver Rusty Wallace, Henry is sports director of WWJ, and has broadcast Detroit Pistons and University of Michigan basketball in his career (From Bob Benko.)

Jeff DeFran is sports director of WTKA-AM in Ann Arbor and co-host of *The Inside Lane* on WJR-AM Sunday evenings, with Jim Mueller. DeFran has experience at several area radio stations and demonstrates versatility as a sports director/reporter/anchor/producer.

Mickey York is co-anchor of the *Detroit Sports Report* weeknights on Fox Sports Net. He came to FSN after four years as sports anchor/reporter at WEYI-TV (NBC), serving the Michigan cities of Flint, Saginaw, and Bay City. He's shown here working the sidelines of the Michigan High School Athletic Association basketball championship in East Lansing (the coach is unknown). He formerly worked for Palace Sports and Entertainment on Detroit Pistons projects. (From Fox Sports Net.)

Damon "The Dog" Perry (left) is host of a mid-morning sports talk show on WXYZ-AM. He's been a sports talk show host previously at WDFN-AM in Detroit. In this photo, he's chatting with one-time WKBD-TV sports reporter/anchor Jerry Millen at Ford Field. (From Ben Eriksson.)

Fox Sports Net correspondent Shireen Saski serves as a feature correspondent, providing reports for various shows including *Red Wings Weekly*, *Tigers Weekly*, *Lions Weekly*, and *Detroit Sports Report*. She also is a sideline reporter on CCHA hockey on FSN. Prior to her Detroit arrival, Saski spent four years as a Dallas-based producer for ESPN. That was preceded by four years at ESPN headquarters in Bristol, CT, from 1991-94, as an associate producer. (From Fox Sports Net.)

Eric Pate is shown interviewing Wayne State University women's basketball coach Gloria Bradley for one of his sports shows. Pate calls play-by-play for WSU basketball and football on WQBH-AM in addition to working various shifts at WDFN-AM. Pate is an accomplished sports writer with Michigan Front Page and Booth newspapers. (From Wayne State University.)

One of the recent additions to the WDIV-TV (Channel 4) sports team is Rob Malcolm, shown here interviewing boxing manager/trainer Emanuel Steward (who is an HBO Sports boxing commentator). Malcolm joined WDIV in the fall of 2001 and has since taken the practice field with the Detroit Lions, laced on a snowboard, strapped on skates alongside an Olympic speed skater, and ridden down a bobsled hill. Before joining Ch. 4, Malcolm was sports anchor at CKVR in Barrie, Ontario. (From Dan Graschuck.)

Jon Bloom (right) loves golf and his listeners love hearing his golf shows on WDFN-AM and WXDX-AM. Shown here with golf professional Phil Michelsen, Bloom hosts *Michigan Golf Live* and *The 19th Hole* on WXDX and WDFN, respectively, during golf season. Twelve stations in Michigan and Ohio carry his *Live* show. The Syracuse University graduate also does play-by-play for Oakland University Grizzlies basketball, and occasionally anchors talk shows and Detroit Pistons broadcasts. (From Tom Waske.)

Jim Stark (left) has been a sports talk show host on WTKA-AM (along with Al Fellhauer), Eastern Michigan University coaches' show host (WTKA), and play-by-play announcer for the Plymouth Whalers on Comcast Cable. He's seen here presenting the Detroit Red Wings/Detroit Sports Broadcasters Association Rookie of the Year award in Nov. 2002 at Joe Louis Arena, to Red Wings forward Pavel Datysyuk. At right is author George Eichorn. (From Ben Eriksson.)

Jeff Lesson has several years of experience at stations WWJ-AM and WXYT-AM, giving sports updates, reporting from various state venues, and hosting his popular *Lesson on Golf* program, in which he gives tips, interviews golf professionals, and answers call-in questions. When not on the air, Lesson practices law (From WXYT-AM.)

As a defensive tackle, Marc Spindler was a battler. Now, he's turned his toughness into often biting comments as a sports talk-show host in Detroit. He's heard co-hosting talk shows on weekends and as a fill-in on WXYT-AM and frequently guests on WDIV-TV's *Sports Final Edition*. This avid hunter and fisherman is from Pennsylvania, but now calls Michigan home since retiring as a Detroit Lion several years ago. (From WXYT-AM.)

Jim Mueller (far right) is another popular motor sports voice in southeastern Michigan. In addition to track announcing duties at Michigan International Speedway—seen here interviewing Muhammad Ali in victory circle—Mueller teams with Jeff DeFran on Sunday evenings for *The Inside Lane* show. Jim has broadcast Cleveland Browns football (over 20 seasons), University of Miami and Louisville University college football, and races at Daytona International Speedway and North Carolina Speedway. (From Michigan International Speedway.)

Doug Karsch has developed a loyal listener base with his talk-show and reporting on WXYT-AM, WWJ-AM and, previously, WTKA-AM. He's as comfortable fielding questions from callers as he is giving his opinions about a variety of sports subjects. An Illinois native, Doug has experience broadcasting University of Michigan sports events and hosting U-M interview programs. (From WXYT-AM.)

Long-time *Oakland Press* sports writer Pat Caputo is another print journalist who moonlights in radio. Caputo was a regular on WDFN-AM before moving to WXYT-AM where is works weekends and on a fill-in basis. His nickname is "The Book" for his vast sports knowledge. He's very opinionated and outspoken during his shows. (From The *Oakland Press*.)

Detroit area sports broadcaster and sports writer Scott Morganroth (right) is seen above interviewing tennis great Jimmy Connors at an event in south Florida. Morganroth has contributed to many stations in Michigan, Arizona, and Florida in his 20-plus years of media experience. He was a regular correspondent for WBRB-AM and WCAR-AM's *The Sports Exchange*. He writes for *The Detroit Monitor* and formerly wrote for *The Hallendale Digest*. Besides Connors, Morganroth has conducted one-on-one interviews with luminaries Muhammad Ali (below), Tommy Lasorda, Al Kaline, Ernie Harwell, Jeff Gordon, Calvin Griffith, and Joe Dumars.

Mark Unger was sports director of WJZZ-FM for many years and is a past president of the Detroit Sports Broadcasters Association (1988-89). He now is a freelance sports reporter, covering Detroit Lions games and other sports.

Milt Wilcox (center) is a talk show guest and occasional host in metro Detroit. He's the former major league baseball pitcher who was within one out of a perfect game while pitching for the Tigers against the Chicago White Sox, before Jerry Hairston broke it up. Wilcox is shown here with an unknown lady and our author George Eichorn after appearing on Eichorn and Phil Guastella's *The Sports Exchange* radio program in 1979 on WCAR-AM.

Jim Hunt is one-half of the University of Michigan Wolverines men's ice hockey broadcast team on WTKA-AM. Hunt is the color analyst while Al Randall does play-by-play. Together, Hunt and Randall broadcast every game—home and away—including post-season games involving U-M such as the NCAA Frozen Four in April 2003 in Buffalo, NY. (From Dave Frisco.)

Earle Robinson has been covering Michigan State Spartans football/basketball/hockey and major pro teams in Detroit for decades. He's employed by WKAR Radio in East Lansing and has hosted *Sports/Talk 870*, a listener call-in show on WKAR. (From WKAR-AM.)

Detroit Sports Report on Fox Sports Net Detroit was hosted in 2001-03 by Marc Soicher and Angie Arlati. Only recently has Soicher been reassigned to host the *Denver Sports Report* by FSN. Marc spent four years as sports director at WCNC-TV in Denver. Prior to that he spent 11 years as weekend sports anchor at WABC-TV in New York City. Soicher also worked TV in Minneapolis, New Orleans, and Columbia, Missouri. (From Fox Sports Net.)

John Ahlers is a another Detroit and Michigan sports broadcaster who has taken his talents elsewhere. Currently the radio voice of the Tampa Bay Lightning of the NHL, Ahlers served in the same capacity for the Detroit Vipers of the International Hockey League during most of the team's existence. He paired with Dave Shand for most games, as heard on WDFN-AM or WXDX-AM. Ahlers also did Vipers' games on Fox Sports Net in Detroit. Ahlers won the Bob Chase Award as the IHL's top broadcaster in 1996. He's worked MSU Spartans hockey, Colorado College hockey, the IHL's Salt Lake City Golden Eagles, and the Louisville IceHawks of the East Coast Hockey League. (From Detroit Vipers.)

Sports Rap co-host Rob Parker (right) interviews Detroit Pistons General Manager Joe Dumars at Ford Field on May 12, 2003, just before Dumars was inducted into the Michigan Sports Hall of Fame. Parker's show airs weekly on WWJ-TV, Channel 62. (From Dan Graschuck.)

One last chuckle from weathercaster and sportscaster wannabe Sonny Eliot (left) as he is greeted by legendary boxing champion Tommy Hearns at the Detroit Sports Broadcasters Association 50th Anniversary dinner in Novi. (From George Eichorn.)

And there are many, many more. Due to space limits we are unable to mention all Detroit area sportscasters or former Michigan residents that have graced the airwaves, but we would like to name the following: Marty Adler, Hank Aguirre, Scott Anderson, Ron Angel, Todd Arner, Sean Baligian, Tom Balog, Jimmy Barrett, John Bell, Steve Bell, Bob Benko, George Benko, Sabrina Black, Mike Bower, Tom Brent, Doug Brown, Jim Brumfield, Bob Buck, Dick Buller, Jimmy Butsicaris, Liz Butsicaris-Jackson, Norm Cash, Dave Chamberlain, Steve Clarke, Bill Collins, Jim Connor, Richard Curbelo, John Cwikla, Chuck Daly, Dennis Davidson, Butch Davis, Joe Donovan, Tony Doucette, Joel Epstein, Mark Faulkner, Roger Faulkner, Ryan Field, Dugan Fife, John Fleury, Jim Forest, John Fossen, Terry Foster, Mike Freedman, Bill Freehan, David Frickman, Dave Frisco, Gary Garman, Dave Geraci, Sal Giacona, Nick Gismondi, Brian Gould, Tim Grand, John Gross, Ron Grubbs, Jerry Hanlon, Dave Harbison, Scott Harrison, Dick Harter, Jim Hendrick, Randy Henry, Bob Hillman, Jason Hillman, George Hope, Don Howe, Fred Huber, Virg Jacques, Red Jamison, Tom Jorgenson, Ed Kaltz, Harry Katapodis, Paul Keels, Ken Kelly, John Kennerly, John King, Quentin King, Stu Klitenic, Tom Kowalski, Pete Krupsky, Bob Kurtz, Lord Layton, Barney Lee, Charlie Lemmex, Mike Lodish, John Long, Paul MacDonald, Dara MacIntosh, Bruce Madej, Skip Maholz, Jay Mariotti, Tom Markowski, Marti Martin, Herman McKalpain, Denny McLain, Jim Measel, Karla Moore, Scott Moore, Terry Moore, John Morales, Rhonda Moss, Jim Mullin, Larry Murphy, Al Muskovito, Bob Newsum, Jim Northrup, Damian Ochab, Dale Ochalik, Tony Ortiz, Gene Osborn, Mel Ott, Rob Otto, John Parker, Charlie Park, Rob Pascoe, Dave Pasch, Marty Pavelich, Fred Pletsch, Dave Quinn, Michael Reghi, Warren Reynolds, Raymond Rolak, Ron Rothstein, Cliff Russell, Rob Sanford, Pete Sark, Tim Saunders, Bob Scheffing, Paul Schneider, Ray Scott, Drew Sharp, Mike Sinnott, Russ Small, Dave Snyder, Lary Sorensen, Tim Staudt, Butch Stearnes, Tom Sullivan, John Tautges, Brandon Tierney, Dizzy Trout, Calvin Usery, Lee Vlisides, Van Vandewalker, Pat Verbeek, Scott Wahle, Dana Wakiji, Jeff Walker, Kevin Wall, John Wangler, Larry Watson, Don Wattrick, Milt Wilcox, Harry Wisner, Bob Wolf, Jon Yinger, and Dan York.